William Benson

Manual of the Science of Colour

On the True Theory of the Colour-Sensations, and the Natural System

William Benson

Manual of the Science of Colour
On the True Theory of the Colour-Sensations, and the Natural System

ISBN/EAN: 9783337145163

Printed in Europe, USA, Canada, Australia, Japan

Cover: Foto ©berggeist007 / pixelio.de

More available books at **www.hansebooks.com**

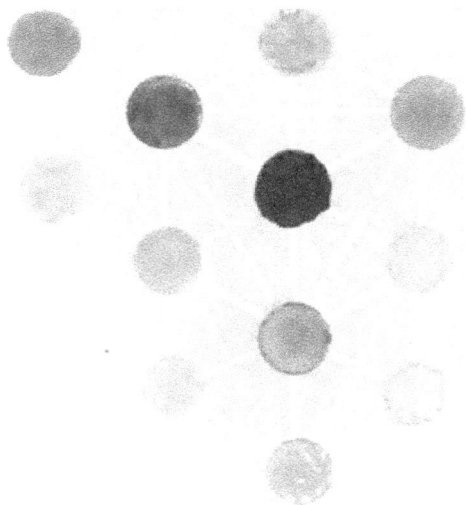

MANUAL

OF THE

SCIENCE OF COLOUR

ON THE

TRUE THEORY OF THE COLOUR-SENSATIONS

AND THE

NATURAL SYSTEM.

WITH A COLOURED FRONTISPIECE AND OTHER ILLUSTRATIONS.

By WILLIAM BENSON, Arch$^{T.}$

AUTHOR OF "PRINCIPLES OF THE SCIENCE OF COLOUR."

LONDON:

CHAPMAN & HALL, 193, PICCADILLY, W.

1871.

LONDON :

VINTON AND SON, PRINTERS, HAMPSTEAD ROAD, N.W.

CONTENTS.

INTRODUCTION.

CHAPTER I.

THE PRINCIPAL COLOURS, AND HOW TO SEE THEM.

CHAPTER II.

THE COLOURS OF OBJECTS SELF-LUMINOUS OR ILLUMINATED.

CHAPTER III.

INTERMEDIATE COLOURS, AND MEANS BETWEEN TWO OR MORE COLOURS.

CHAPTER IV.

THE NATURAL SYSTEM OF COLOURS AND ITS USE.

CHAPTER V.

OCULAR MODIFICATIONS OF COLOURS; THE SENSIBILITY OF THE EYE.

CHAPTER VI.

THE HARMONY OF COLOUR.

CHAPTER VII.

MENTAL EFFECTS OF COLOURS AND COLOUR-COMPOSITIONS.

CHAPTER VIII.

PECULIAR AND DEFECTIVE COLOUR-VISION.

ILLUSTRATIONS.

INTRODUCTION.

THE Science of Colour teaches the nature and causes of colours; their distinctions; their relations to each other; their classification; the mental effects that attend them, and the causes and laws of their harmony. It includes the modifications of colour arising from the varying sensibility of the eye, and also the peculiarities of colour-vision which are found to exist in different persons. It enables the eye to distinguish and recognize the different hues in their various tints and shades, and to acquire a correct judgment about the real constituents of colours, upon which their effect in any composition depends; it improves and directs the natural taste, and aids the student in devising as well as appreciating good compositions of colour.

The whole subject is full of interest and beauty, and an acquaintance with it adds extremely to the pleasure to be derived from contemplating natural objects and scenery, as well as pictures and other works of art; and it has one advantage, perhaps, above any other branch of natural philosophy; it teaches us in a very remarkable way to distinguish between sensations and their causes, and not to judge hastily according to the appearance of things. And as the chief difficulty of the science seems to arise from this, it is proper in the Introduction to call the student's attention especially to it, and to guard against those misconceptions which might otherwise hinder his progress.

First, then, the science teaches us that colours are purely and simply sensations, excited by the action of light on the nervous tissue of the retina which forms a screen at the back of the eye; and a little observation and reflection is enough to convince us that they must be so, since they are only seen by the aid of light, and vary with different kinds of light. But when we look at an object, we see so clearly the colour on its surface, at a distance from us, and that colour under the ordinary light of day is so constantly the same, that we can hardly imagine it anything but some inherent quality of the surface, belonging to the object, just as much as the form which we see. This is therefore the notion which unreflecting persons usually have about colours, insomuch that it is not always easy to make them understand the difference between colours (properly so called) and the pigments or colouring matters which go by the same names.

But suppose one has advanced beyond this primitive stage, and knows that colours are caused by the light that enters the eye; and suppose one has learnt also that there are innumerable different kinds of light, each distinguished by some difference of colour, and that a difference of colour in two objects certainly indicates some difference in the kinds of light which they send to the eye; what then is more natural than to infer that colour is an indication of some distinctive quality residing in the light? Even eminent philosophers have hastily assumed it to be so; and some have supposed that there must be as many different species of light as there are distinct elementary sensations of colour, which is inconsistent with what is known of the nature of light. And it is still more common for those who have not studied the science of colour, to think that the colours of simple or homogeneous light (light that is all of one kind) are themselves simple or homogeneous. Yet our science tells us that all the colours of homogeneous lights may be more or less compound sensations, and most of them certainly are so; and that mixtures of different kinds of light may have the same colours. For the colours of even certain kinds of homogeneous light may be imitated by mixtures of other kinds; and again the very same kinds of light may have different colours, not only in different eyes, but also under different circumstances in the same eye.

Hence we must adhere, contrary to all natural prejudices, to the doctrine that colours are sensations, simple or compound; and it must be clearly understood that when their names are used as adjectives to describe any light, as red light, yellow light, white light, they mean only that the light, whatever it is, has the property of exciting in our eyes the sensation of the colour red, yellow, white, or whatever it may be. In like manner, when the same terms are applied to pigments or other material objects, as red paint, yellow solution, white paper, they must be understood to mean that the lights which the objects send to the eye, under the ordinary illumination of daylight, produce those same sensations.

Again, our science teaches that some colours, such as whit₁ and yellow, are compound sensations. A perfect white, for example, it tells us, is a combination or mixture of three distinct sensations, namely, red, green, and blue, excited at once in equal force on the same part of the retina. A perfect yellow, it tells us again, is a combination or mixture of two of these, red and green, in equal force. Yet, when we contemplate a white object, we perceive nothing whatever to call up the idea of red, or of green, or of blue, or of any other colour. It seems perfectly homogeneous, simple and pure. And when we look at a yellow surface, it is by no means easy (though perhaps not quite impossible under certain circumstances) to imagine we see a combination or mixture of red and green. Nevertheless, a careful investigation of the matter shows that these really are compound sensations as asserted. For we can compound white light by throwing together or mixing red, green, and blue lights, and yellow light by throwing together or mixing red and green lights.* We can also show that the light given out by the best white and yellow objects contains a mixture of the same components†. And we can produce the sensations of white and yellow by mixing the sensations of red, green, and blue together, and those of red and green together, without mixing the lights.‡ Hence, strange as it seems, it must be admitted that sensations apparently simple may really be compound; and their components such as cannot be recognized in them.

To mention another instance, it is well known that if we mix together two coloured solutions, or two transparent pigments, or superpose two coloured glasses, against a white surface, we produce a third colour. What more natural than to suppose that this new colour is a combination or mixture of the colours of the two solutions, pigments, or glasses separately? Yet, our science says this is only approximately true in certain

---

* By mixing opaque coloured powders; by reflection from glass (as in Sect. 36); and by actually throwing together beams of light that have passed through coloured glasses.

† By analysis with the prism (Sect. 34); by analysis with coloured glasses (Sect. 22).

‡ This is evidently effected when colours are blended by the method of rotation (Sect. 36); or by blending two separate sensations excited in the two eyes, as when we see double images of two different coloured spots, one of each colour overlapping.

cases; in most cases the third colour must differ widely from the colour that would result from the mixture of the two colours. Thus, green has been commonly supposed to be a mixture of yellow and blue, because it is easily produced by a mixture of yellow and blue pigments; yet it is easy to show that the true mixture of those two colours is not green but gray; and it is easy also to explain why green is produced by the mixture of such pigments.

It is equally natural to suppose that by making a coloured solution stronger or thicker, or by laying on a thicker wash of a pigment, we increase the strength or intensity of the colour, without other alteration. It has even been proposed to use a combination of hollow glass wedges, filled with coloured solutions, as a means of measuring the intensity of a colour according to the thicknesses of the solutions which give a colour to match it; and experiments made in this way are the basis of what is commonly called Field's doctrine of chromatic equivalents. Yet the truth is easily shown that to thicken a solution or transparent pigment materially alters the hue as well as darkens its colour; and that it is absurd in any case to make the thickness of a solution the measure of the strength of its colour.*

Once more, as we notice the brilliant whiteness of the sun, and of other intensely luminous bodies, and also the whiteness of many illuminated objects, as clouds, snow, linen, paper, chalk, far brighter than other colours, and the white superficial reflection that takes place from even dark-coloured objects, quite independently of their proper colours; and as we find, too, that a white illumination is necessary to bring out all the various hues of objects, we easily connect the ideas of whiteness and of light, and infer that other colours are something essentially different from white, and perhaps think, as some have done, that they rather partake of the nature of darkness. Other phenomena added to these, such as the blueness of the light reflected from very small reflecting particles in a transparent medium (which is only noticed when viewed against a dark background), and the prevailing redness of the

---

* In truth colouring matters act by destroying some kinds of light faster than others; so that the lights that escape from thick washes of a transparent pigment cannot be the same in quantity or quality; and the light that escapes from two pigments is no more a mixture of the lights that escape from both of them separately, than the grain which falls through both of two different sieves is a mixture of that which falls through each one of them alone (Sections 20-25).

light which is transmitted through most bodies imperfectly transparent, have led, not only ancient philosophers, but also ingenious moderns, like Goethe, to suppose that bluish colours are constituted by darkness seen through a white or light medium, and reddish colours by white or light seen through a dark medium.

But our science utterly contradicts all such notions, constraining us to believe that white only differs from other colours in containing all the simple colour-sensations in equal strength. It makes no difficulty in explaining the remarkable phenomena alluded to above; showing us on the one hand that bodies just heated enough to be luminous give out red light, and when more heated give out other kinds also, and all more and more, so that their light becomes not only brighter, but also whiter; and on the other hand, that bodies which reflect all kinds of light equally, often reflect all very powerfully, so that white objects are commonly bright; whilst the blueness of the sky, and the redness of the sun through a fog, and thousands of other effects, are necessary consequences of the various ways in which matter acts upon light.

Whenever, therefore, the term "light" is used (as it often is) to express a sensation, it must be understood as a general term, including each and all of the colour-sensations, and not as denoting any other sensation excited by the action of light, for there is none beside red, green, blue, and their compounds.

The importance of understanding these points is great; since whole theories and systems of colour have been founded upon erroneous conclusions hastily drawn from natural phenomena, correctly observed, but not understood; and it is not easy to deduce reliable and useful rules in colour-designing from such defective principles. Besides this, a most beautiful branch of natural philosophy has been neglected and discredited in consequence of them. Through the rashness of several esteemed writers in advancing crude speculations about colour, and through ignorance of what has been really discovered, many are disinclined to entertain the subject at all. They either think that nothing certain is known about it, or that all theories will prove equally useless in art; or else are wedded to preconceived opinions which they are unwilling to disturb.

Hence a manual which aims at freeing the subject from previous errors and imperfections, should call special attention to the proofs of its doctrines (so far as they are not universally accepted), and the authorities that support them, as well as the advantages which will attend their adoption.

Now it may be safely asserted that by a little intelligent study of the facts revealed in those simple and beautiful experiments which are described in this treatise,* any person who has the use of his eyes, and believes the grand propositions about light which Newton established when he first unfolded to the world the meaning of the prismatic spectrum, may very quickly satisfy himself of the truth of the doctrine. And it may be safely asserted, too, that when the matter is once understood, no one who takes interest in the harmony of colour will feel any wish to revert from light to darkness, or regret the few hours of interesting study which enabled him to master a charming science, hitherto generally neglected by those who should profit most by it.

But accurate researches for determining the colour-sensations excited by the prismatic rays are not wanting; and they fully establish the doctrine. Those recently made in this country by Mr. James Clerk Maxwell have just now been confirmed by others made on another plan by Herr J. J. Müller in Germany.† Experiments by Sir John F. W. Herschel, and by Professor Helmholtz, might also be cited in confirmation. For these investigations have distinctly proved that both white and all possible colours can be produced by mixtures of the best prismatic red, green, and blue, and by mixtures of no other three colours but these; quite contrary to the common red-yellow-blue theory. But that this common theory is utterly wrong is obvious from the simplest experiments, which prove that yellow consists, as before mentioned, of the sensations of red and green excited together, and is exactly complementary to blue. So that if red and blue are simple colours, yellow must be compound.‡

---

* See especially Sections 0-13, 34, 35, 36, 38, 39.

† Transactions of Royal Society, 1860, and Poggendorff's Annalen, March and April 1870.

‡ The true doctrine of the three colour-sensations was clearly stated at the beginning of this century by that great philosopher Thomas Young (Natural Philosophy, Lecture xxxvii.), and has lately been explained in a popular manner by Sir John F. W. Herschel (Essay on Light in "Good Words" for 1865, which has also been reprinted in his Familiar Lectures), as well as by Mr. Maxwell (Lecture at the Royal Institution, 17 May, 1861). Notwithstanding the seeming difficulty of reconciling it with first appearances, the neglect it has hitherto met from artists is strange, so great and so evident is its superiority, for all the purposes of art, over the conventional system that has been substituted for it.

In several letters to the " Builder " (in the latter part of 1865), and in a paper read at the Institute of Architects (Sess. Paper No. 7, 1869), the author himself has endeavoured to expound the proofs and uses of the science of colour. Since then, he

Another advantage of learning the true nature and relations of colours will be found in the improvement to be derived from the intelligent study of the rich compositions of colour presented in natural scenery and natural objects, and also in works of art. This will greatly increase the value of those illustrated works on colour-decoration and ornament which have been so abundantly produced of late years. For those who would study these to profit, must not only feel that some combinations of colour are preferable to others; they must also know the reasons why; otherwise they can but make empirical attempts to improve. Now a false theory mystifies the matter, and even if sound rules are hit upon, often hinders the right application of them; but true knowledge will ever improve the judgment, and direct the hand; and if "a thing of beauty is a joy for ever," it is doubly so when the understanding co-operates with the feeling.

The nature of light itself is perhaps not strictly a part of the science of colour; but it adds much to the interest of the subject to know that the wonderful medium through which we have communication with the utmost parts of the universe, the all-pervading and perfectly elastic ether, receives and transmits in waves of vibration, with inconceivable velocity in every direction, all sorts of shocks, however minute, arising amongst atoms of matter, just as air does with those arising from the tap of a hammer or the ring of a bell; and that these etherial waves, when they are of the kind called waves of transverse vibration, and their time or period falls within certain limits, constitute light;—the different kinds of light differing essentially in their wave-times alone.

has met with a treatise on colour, prepared at the suggestion of the Directors of the Imperial Austrian Museum of Art and Manufacture, by Ernest Brücke, Professor of Physiology in the University of Vienna, and which has been translated into French (Des couleurs au point de vue physique, physiologique, artistique et industriel, par le Dr. Ernest Brücke. Traduit d'allemand par J. Schutzenberger, Paris, 1865). The progress of the science of colour on the continent appears from the fact that in this work the whole subject is treated in a scientific manner, and what is wrong in the common theories is rejected, though no complete system of colour is proposed. Field's doctrine of chromatic equivalents, which has long held too much authority in England, Professor Brücke declares "is false from beginning to end, and owes its origin to an incorrect interpretation of natural phenomena." Sir J. F. W. Herschel, after his life-long labours for science, may be allowed to speak with authority here. In a letter to the author, he tersely describes the prevailing opinions about light and colours as "idola, which, from a jargon, have fixed themselves into a doctrine." It may be long before that doctrine loses its hold on the public; in lectures, and even in new treatises on Natural Philosophy, its truth is still sometimes assumed by persons who ought to know better; but at least those who direct the education of students in art should take care not to fall behind the age, or allow errors repudiated in Germany and France to remain unchallenged in England.

Reflection and refraction, with all their complicated and beautiful phenomena, are but natural consequences of such waves meeting with different media—portions of space in which the etherial density and elasticity are modified, so that the waves traverse them with altered velocity. The equally marvellous phenomena of interference, which led to the discovery of the nature of light, and the actual measurement of the lengths of the luminous waves, arise from the action of one wave upon another, when two either concur to increase the disturbance, or oppose each other to quiet it.

How the minute wave-lengths, and still more inconceivably short wave-times, of the different kinds of light have been measured, and what the results of the measurements are, must be left to works on optics to recount. It is enough to note here that their periods are all so short that there is time, even in the fraction of a second during which the sensation excited by a luminous wave continues, for a succession of innumerable separate waves of any kinds of light to follow each other along the same line or ray; so that we may have lights of all kinds mingled together in any proportions, and making any conceivable mixture of their separate colours at the same moment on the same part of the retina.

The student should of course be so far acquainted with the action of the eye, as to know that the light which proceeds from the various objects in the field of view is caused by it to fall regularly upon corresponding parts of the retina, so as to form upon it a picture of the whole scene, like that produced in a *camera obscura*; and that when the direction of the eye is changed, this picture shifts its position, so that the part of the sensitive screen which was before affected by the light from one object, is now affected by the light from another, or perhaps by none at all. It should be noticed, too, that the sensations excited on corresponding parts of the retinas of the two eyes become blended together by the union of the two branches of the optic nerve.

MANUAL

OF

THE SCIENCE OF COLOUR.

CHAPTER I.

THE PRINCIPAL COLOURS, AND HOW TO SEE THEM.

1. *Colour-sensations, simple and compound.* The eye is so constituted in the great majority of persons as to be capable of three simple or elementary sensations of colour, which are best described by the terms red, green, and blue.*

The simple sensations are never excited separately, with perfect purity, on any part of the retina, but always accompany each other in greater or less degree, causing the endless variety of colour which we observe.

When all are excited at once with equal intensity on the same part of the retina, the result is the sensation of white.

When one of them is excited in less degree than the others, the two predominant sensations form a compound or mixed sensation, accompanied with a mixture of white proportionate to the degree in which the third is excited. These binary compounds are—

(1) Seagreen, which consists of green and blue in equal strength, and passes through seagreen-green into green, when the blue is weakened, and through seagreen-blue or azure into blue, when the red is weakened.

---

* As these and all other names of colours are commonly applied to many different hues, it should be observed that the particular hues here intended are those of the deepest prismatic red, green, and blue, as will be explained further on. Cases of peculiar and of defective colour-vision are noticed at the close of this treatise.

(2) Pink, which consists of blue and red in equal strength, and passes through pink-blue or puce into blue, when the red is weakened, and through pink-red or cerise into red, when the blue is weakened.

(3) Yellow, which consists of red and green in equal strength, and passes through yellow-red or orange into red, when the green is weakened, and through yellow-green into green, when the red is weakened.

When two of the simple sensations are equally excited in less degree than the third, we have of course the predominant sensation, together with a mixture of white proportionate to the degree in which the other two are excited.

The simple sensations are called primary colours, or primaries; and their binary compounds, secondary colours, or secondaries.

White, the peerless ternary sensation, though in common language sometimes distinguished from colours, is in truth the chief of all. Black, on the other hand, though no positive colour at all, must yet for convenience sake be included amongst them, as the lowest limit of all colours.

2. *Hues, tints, and shades of colour.* The absolute excess or predominance of some one or two of the simple sensations over the rest, as above described, is called the hue of a colour; and the greater such excess, the stronger the hue.

The whole series of hues is best remembered by conceiving the six gradations from red to yellow, from yellow to green, from green to seagreen, from seagreen to blue, from blue to pink, and from pink back again to red, arranged in order and in equal strength on the sides of a hexagon, as indicated in the outer row of coloured spots in each of the diagrams on the frontispiece. A similar arrangement is frequently described under the name of the chromatic circle; but the hexagon is the preferable form, because it distinguishes the six points from which new gradations start. White and black, being perfectly neutral, or without any hue, form no part of any chromatic circle.

If the deficiency of each sensation in a colour is diminished in proportion, until all the three are brought to the full and equal intensity which they possess in white, the colour changes by a regular gradation into white, without any variation of hue. The lighter colours thus produced are often called tints of their respective hues.

The brightness of a colour is the sum of the intensities of all the simple sensations of which it consists. The proportionate diminution of these from the most powerful that the eye is capable of down to the absence of all sensation, produce all the shades of colours, from the brightest down to black.

3. *Excitement of colour-sensations.* Colour-sensations may be slightly and promiscuously excited by various means, as by pressure or an electric shock, and even without any external application, as seen in the undefined images that seem to float before the eye in the dark. But the action of light from without is incomparably the most powerful and effective means of producing them ; and it is for the admission of light in proper quantities, and its regular distribution on the surface of the retina, according to the direction from which it comes, that the transparent parts of the eye are designed.

4. *Cause of variety of hue.* Each of the innumerable different kinds of light, in each degree of intensity, produces a peculiar mixture of the three simple sensations of colour, one or two always predominating more or less over the other.

When two or more of the different kinds of light enter the eye together, or in very rapid succession, they produce an effect compounded of the effects which the separate lights would produce; in other words they make us see a colour compounded of the separate colours which we should have seen had the same lights come separately.

5. *Separation of mixed homogeneous lights.* In consequence of the rays of each kind of light being differently bent or refracted in passing from one trans-

parent substance into another, a ray of mixed light going through the angle of a prism is always divided into the several kinds of light of which it consists; and these, being all bent at different angles from their original direction, when they enter the eye fall upon different parts of the retina, and excite each its proper sensation of colour apart from the rest. Thus is produced the splendid series of colours known as the prismatic colours, and seen in the solar spectrum, or, with less purity, in the rainbow; and these are the colours of the several different kinds of light, arranged in order of the refrangibility of the lights that produce them. A knowledge of them is of great importance, and can only be obtained by actual observation. For they far surpass the best representations that can be made by pigments or dyes.

6. *What light gives a complete spectrum.* An uninterrupted spectrum, exhibiting a complete series of the prismatic colours in sufficient intensity to make them all distinct, can only be obtained from the brilliant white light given out by any intensely heated solid, such as the lime in the oxy-hydrogen light, or the carbon in the electric light.

Such, for all purposes of the study of colour, is the light of the sun, either direct, or reflected from a bright white surface; though when accurately examined its spectrum is found to be interrupted by innumerable narrow dark lines, indicating rays which are wanting, having doubtless been absorbed or lost in traversing the solar atmosphere or our own.

7. *How to obtain the spectrum.* To obtain a spectrum sufficiently brilliant and pure for studying the colours, the readiest way is to view a bright and narrow white stripe through a good prism. The stripe may be a portion of a bright white cloud seen through a slit in a dark screen, or it may be a piece of silver wire, or a slip of white paper, laid across a dark cavity, and brightly illuminated. The more brilliant it is, the narrower it may be, or the greater the distance at which it may be viewed, without too much diminishing the

light; and therefore the more nearly pure the spectrum will be.

Fig. 1 shows how the prism may best be placed to view the spectrum of the light of a white sky through a horizontal slit in a shutter.

Fig. 2 shows how it may be placed to view a white stripe on a dark ground.

The dotted lines converging to the eye in the direction of the rays emerging from the prism are intended to indicate more distinctly the direction in which the spectrum will appear.

The refracting angle of the prism should always be parallel to the bright line, and should be so directed that the light may enter and emerge at nearly equal angles with the sides of the prism, as indicated in the figures. This position deflects the light as little as possible from its original direction. It is easily found by slowly turning the prism one way or the other on an axis parallel to its edge, till the spectrum becomes stationary.

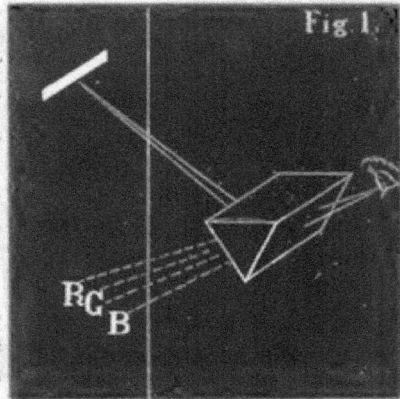

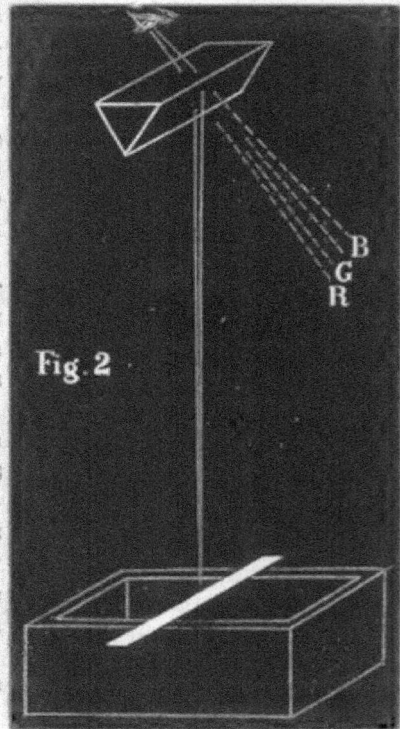

When a sunbeam enters a dark room through a slit in a shutter, and passes through a prism held parallel to the slit, the brilliancy of the spectrum (too great for direct vision) allows it to be received on a white screen. In this way its splendid colours may be seen by many spectators at once, the white surface reflecting it in every direction.

8. *The prismatic colours.* On the first view, a good spectrum appears to consist of three bands of colour, red, green, and blue: the first including all the less refrangible rays, the second all those of mean refrangibility, and the third all the more refrangible rays. But on a closer inspection it appears that the colours change gradually into each other, and the principal varieties are stated in the following list :—

Red band
{
Dark Red.
Bright Red or Scarlet.
Yellow-red or Orange.
}

Green band
{
Yellow.
Yellow-green.
Green.
Seagreen-green.
}

Blue band
{
Seagreen.
Seagreen-blue or Azure.
Blue.
Dark Blue.
Very dark Violet.
}

The last of these is scarcely visible unless the spectrum is produced from very strong light. It should be noticed that the rays of the red band are most condensed, and those of the blue band least condensed, because the glass disperses or spreads out the former much less than the latter. This makes the red band narrower and brighter, and the blue band wider and darker, than they otherwise would be. The brightest part of the spectrum is between the yellow and green ; but if all the rays were equally dispersed it would be about the middle of the green band.

9. *What are the primary colours, and how to see them.* The best prismatic red, green, and blue, near the beginning, middle, and end of the spectrum, are so much deeper (or stronger in hue in proportion to their brightness) than the colours which lie between them in the spectrum, that by their mixture all the latter may be produced in the full strength in which they are found in the spectrum, whilst they themselves cannot be produced by the mixture of any other rays. Now the colours of all objects must be mixtures of prismatic colours; therefore, since these are compounded of the prismatic red, green, and blue, it follows that all possible colours are so compounded, and that the prismatic red, green, and blue give the nearest possible approach to the three simple sensations of colour. For the sake of shortness therefore these may be called the primary colours.

To see them in their greatest strength, we should look through the prism at a white band of a moderate breadth upon a black ground; the colours so produced are much stronger than those of the pure prismatic rays, produced by looking at a very narrow line of the same white, because numbers of those rays very nearly resembling each other in colour are thereby thrown together. It is easy to adjust the distance of the prism so as to produce the strongest green which is possible out of the light that comes from the white band, and the same position of the prism will give also the strongest red, and very nearly the strongest blue.*

10. *What are the secondary colours, and how to see them.* When we view in like manner through the prism a black band upon a white ground, we obtain a spectrum that is exactly complementary to the spectrum of a white band of the same width on a black ground. In other words, the colour of any part of the one spectrum contains exactly what the colour of the

* Owing to the blue rays being more dispersed than the green, the distance of the prism from the object should be rather less to throw together all the blue rays.

corresponding part of the other spectrum wants, to make up the full white. For as the white band would exactly cover the black band, and reduce the whole to a uniform white, so would the spectrum of the white band cover that of the black, and every colour in it would reduce the corresponding colour in the other to white. If therefore the one exhibits the three primary colours in greatest strength, the other must give, opposite to each, a combination of the other two primaries in greatest strength. These may therefore be called the secondary colours.

The seagreen which we thus obtain is produced by throwing together all the prismatic rays of the blue and green band without those of the red band; the pink, by throwing together all the red and blue, without the green band; the yellow, by throwing together all the red and green, without the blue band. Each has a brightness equal to the sum of both the primaries by which it is constituted.

11. *How to see the primaries and secondaries in contrast.* The primary and secondary spectra in the two preceding sections are best seen side by side, by making the black and white bands join, as in Fig. 3. By this means the complementary colours are brought into juxtaposition, and by the aid of the prism the eye instantly perceives the true complementary to any one of the splendid series of colours on the right or on the left of the central line. Care should be taken to hold the prism parallel to the white and black bands, and opposite to the centre of the diagram, that the complementary bands may join end to end without confusion.

12. *How to see some of the best combinations of the prismatic colours in contrast.* By viewing through a prism an edge of white against black in a line with an edge of black against white, as in fig. 4, the beautiful series of colours produced by throwing together larger and larger continuous parcels of the prismatic rays, beginning from the red end of the spectrum, may be seen in juxtaposition with their perfect complementaries, formed by throwing together smaller and smaller parcels

of the rays beginning from the violet end. These colours, like those mentioned in the last section, deserve careful study. The following lists show how the two complementary series are constituted :—

$R + G + B$	$= White$	0		$= Black$
$R + G + \frac{B}{2}$	$= Light Yellow$	$\frac{B}{2}$		$= Dark Blue$
$R + G$	$= Yellow$	B		$= Blue$
$R + \frac{G}{2}$	$= Yellow\text{-}red$	$B + \frac{G}{2}$		$= Seagreen\text{-}blue$
R	$= Red$	$B + G$		$= Seagreen$
$\frac{R}{2}$	$= Dark Red$	$B + G + \frac{R}{2}$		$= Light Seagreen$
0	$= Black$	$B + G + R$		$= White$

The middle colour on each side is about half as bright as the full white.

13. *How to see at once all the best combinations of prismatic colours.* Another beautiful and instructive experiment with the prism exhibits at one view all possible combinations of continuous parcels of the prismatic rays. This is effected by viewing through the prism a triangular spot of white upon a black ground, and a similar spot of black upon a white ground as in fig. 5, the prism being held at such a distance that small spots of white and black shall appear respectively on the left hand edges of the two spectra.

The first of these spectra being formed by the partial overlapping of an infinite number of triangles of all the prismatic colours, unequally displaced by the prism, must contain the colours of all possible combinations of the prismatic rays in single parcels, large or small; while the other must contain in each corresponding part the complementary colour proper to the combination of all the remaining rays.

14. *Value of such observations.* In the experiments mentioned in the last three sections the opposite colours are perfect complementaries; each pair when thrown together making the white from which the spectra are produced. It follows therefore

(1) That every colour seen in these experiments equals its opposite colour in strength of hue.

(2) That if any of the colours exceed the mean brightness (that is, half the brightness of the white), its opposite colour falls as much below it.

(3) And therefore that the depth of each of the colours is perfectly matched by the clearness of its opposite colour—the depth of a colour depending on the proportion which its strength of hue bears to its brightness, and the clearness of a colour depending upon the proportion which its strength of hue bears to its darkness, or defect from the brightness of the white.

Thus, the study of the colours exhibited in these experiments, is not only valuable for teaching us the colours that are perfectly complementary in hue and in brightness, but also what is perhaps quite as important, the colours that match in depth and clearness.

15. *The best standard colours.* The prismatic colours and their regular compounds produced in these experiments, are far superior in depth and clearness to the best colours of pigments and other natural objects resembling them in hue. The cause of this will be shown in the next chapter. They therefore form better, as well as more certain and uniform and more easily available standards of colour than can otherwise be obtained. Thus, the colours of the best collections of the red, green, and blue rays may be taken as the standard primaries, and the seagreen, pink, and yellow produced by throwing these together in pairs, may be taken as the standard secondaries. Still it must be remembered that even these are but approximations to the unattainable simple and binary compound sensations of colour, which, together with perfect white and black, constitute the eight principal colours.

CHAPTER II.

THE COLOURS OF OBJECTS, SELF-LUMINOUS OR ILLUMINATED.

16. *Colours of incandescent solids.* All bodies, not in the state of gas or vapour, when so heated as to begin to be luminous, emit first the extreme red rays of the spectrum, and as their heat increases emit rays of higher and higher refrangibility. Hence their colour is at first red, as seen in iron beginning to glow ; then orange, bright yellow, and still brighter pale yellow, as seen in iron more strongly heated, in burning coals, or in flames in which a stream of incandescent particles of solid carbon is continually pouring forth ; and lastly (when the substance is so strongly heated that it emits all kinds of light abundantly) it becomes a dazzling white, as seen in burning magnesium, or in the lime in the oxy-hydrogen flame, or in the carbon of the electric light.

17. *Colours of incandescent vapours.* Gaseous substances, or vapours, in a state of incandescence, emit only some peculiar kinds of light ; so that each is distinguished by its own colour, and their light when analysed by the prism presents a spectrum of one or more bright lines alone. Thus the light given out by burning vapour of alcohol or spirits of wine is a pale blue, but if common salt is put on the wick, the incandescent vapour of the salt gives out in addition a bright yellow light which belongs entirely to a narrow line in the

yellow part of the spectrum, and is distinctive of the metal sodium contained in the salt.

In like manner potassium imparts a purple light, compounded of red and blue rays; lithium a brilliant red light; copper a green light. Several such substances, which are easily vaporized, and whose glowing vapours are remarkable for their colours, are used to make coloured flames and fireworks.

18. *Effects of different kinds of illumination.* Objects which are not self-luminous can only be seen by that portion of the light of some luminous body which they reflect or transmit to the eye. Their colour therefore depends (1) on the kinds of light with which they are illuminated, and (2) on the proportions in which they send those kinds of light to the eye; for while one body may send to the eye all kinds of incident light in the same proportion, another (having the property of reflecting or transmitting some kinds of light more freely than others) sends them in different proportions.

Bodies which send to the eye all kinds of light in like proportion, always appear of the same sort of colour as the light which illuminates them : thus, for instance, they are white by daylight, yellow by candlelight, red by a red light; their brightness depending, of course, on the quantity of light which they send to the eye.

But bodies which send to the eye a larger proportion of some kinds of light than of others, often differ widely from the colour of the incident light. Thus things appear of different colours, though illuminated with the same white light of the sun ; some sending red rays to the eyes in greatest proportion, others green, and so on.

Such objects also may have a strong hue by one sort of illumination and be almost black by another; as things deep blue by daylight seem black by candlelight, from its deficiency in deep blue light. Or again, they may change their colour remarkably ; as what is seagreen or pink by daylight appears green or red by candlelight, and what is white by daylight appears yellow by candlelight, all from the same cause. Hence also objects may sometimes appear of different colours under one kind of illumination, and of the same colour under another, as those which are white and pale yellow by daylight, appear both of one yellow by candlelight.

19. *Causes of colours of non-luminous bodies.* When light falls on the surface of any object, part of it enters the body, and part is reflected without entering. Of that part which enters, part is absorbed or lost within the body, and the remainder is either transmitted through the body and emerges on the other side, or is reflected back again from within the body, and emerges on the side on which it entered.

The endlessly varied colours and appearances of non-luminous bodies arise from the different degrees and ways in which they thus affect the light that illuminates them. In noticing the principal varieties, it will be supposed, unless the contrary is expressly stated, that the illuminating light is white and consists of a mixture of all kinds of light.

20. *Colours from transmission of light.* First, colours are produced by light transmitted through an object.

These depend on the quantities of the different kinds of incident light which escape reflection both from the outer surfaces of the body and from within it, and also escape absorption in the body. The power of reflection being generally the same for all kinds of light, differences of colour mainly arise from peculiar powers of absorption. Some bodies, like air, water, or clear white glass, absorb no kind of light sufficiently to make a noticeable alteration in the colour when the thickness is moderate.* Others, like a solution of verdigris, absorb the red rays most, and therefore appear seagreen; others absorb the green most, and therefore appear pink; others the blue, and therefore

---

* But even the very clearest of these will be found to have their peculiar colour when the thickness is sufficiently increased; for there is no ponderable substance which does not absorb light at all. That the atmosphere obstructs the passage of all kinds, and of some more than others, is evidenced by the dimness and redness of the setting sun, seen through a thickness of many hundred miles.

appear yellow. Others again absorb equally both the green and blue, and appear red ; or absorb equally the blue and red, and appear green ; or absorb equally the red and the green, and appear blue. Lastly, some absorb nearly all the rays with such great power that, like black glass or common ink, they are almost totally opaque except in small thicknesses.

Two varieties of bodies giving colour by transmitted light should be noticed :—(1) Such as transmit light regularly, which, as they allow the forms of bodies to be seen through them, are called transparent. (2) Such as transmit it irregularly, scattering it more or less by internal reflection, and are therefore only translucent.

In order to see the colour of any such translucent body by transmitted light, it is necessary to view it against a white sky, or other sufficiently bright white surface, no light being allowed to fall on the side next to the eye.

21. *Effect of increasing the thickness of a translucent body.* Owing to the difference of the rates at which bodies absorb the different kinds of light, the colours of translucent bodies usually vary with the thickness of the body. Thus a solution of potassium bichromate gives first a pale yellow; in greater thickness, a golden yellow; then orange, approaching at last to dark red. One of copper sulphate varies in like manner from bluish seagreen to dark blue. But those substances which almost destroy at once all the prismatic rays except a few to which they are very transparent, do not perceptibly change their colour except by becoming darker, when their thickness is increased ; as may be seen by looking through one, two, three, or more thicknesses of the common red glass which owes its colour to copper oxide, and is very opaque to all but some of the red rays.

The colour of a translucent body generally approaches nearer to some particular prismatic colour as the thickness of the body increases ; namely, the colour of that kind of light which traverses its substance most freely.

In that case, therefore, they usually become deeper and more striking to the eye.*

22. *Effect of superposing translucent bodies.* On these principles it will be easily seen that the colour of two such bodies superposed must not be regarded as a mixture of their separate colours, but as merely the colour of those prismatic rays which happen to traverse both the bodies. Thus a white sky, viewed through a seagreen glass and a yellow glass laid together, gives a good green, but much darker than the seagreen or the yellow of the separate glasses; for each of these glasses allows the green light to pass through, whilst the red component of the white light is destroyed by the seagreen glass, and the blue component by the yellow glass. In like manner, pink and yellow glasses together give a red colour, and pink and seagreen glasses a blue colour.

The composition of colours may often be shown thus. If a yellow object, for instance, appears as bright through a red glass, or through a green glass, as a white one does, but appears black through a blue glass, it shows that the light of the yellow object has in its composition all the red and all the green, but none of the blue light of the white object.

23. *Colours from internal reflection.* Secondly, colours are produced by light reflected from within the substance of the object.

This is the commonest case, including the largest number of organic and inorganic substances. A large part of the illuminating light enters the body, and after traversing some small depth of its substance, the part of it which escapes absorption is reflected; the part of this reflected light which again escapes absorp-

* In consequence of this, some have supposed the strength of the colour is increased by increasing the thickness; as if there were more blue, for instance, in the colour of a thick solution of copper sulphate than in that of a thin one. But it must be remembered that the absolute quantity of blue is greater in the colour of the thin solution, though it is there combined with a quantity of green, from rays which are lost in the thick solution.

tion on its return, emerges to enter the eye. Thus these colours, like those of translucent bodies, are caused by the absorption of light. Indeed those pigments and dyes which give their colour when laid over white substances, may be regarded almost entirely as translucent substances, since their colour is mainly due to light that passes through them to the white reflecting substance beneath, and then returns through them again to the eye, the remainder of the incident light being extinguished within the substance. In these cases the stronger the reflecting white, and the more transparent the pigment or dye, the brighter the resulting colour; as seen in the brilliant effects produced by transparent varnishes laid over burnished metal.

24. *Effect of mixing pigments.* When bodies of this kind, such as soluble or finely-ground pigments, are mixed together or laid in washes one over another, their combined effect is like that of superposed glasses of the same colours; but the more nearly the pigments approach to the condition of opaque powders, the less the resemblance. Thus, a mixture of Prussian blue and gamboge produce a very good green; whilst cobalt blue and chrome yellow powders, which are more opaque, give a colour approaching to gray, which is more truly the mixture of their separate colours.

25. *Effect of thickening a pigment.* The colours given by such substances in very thin layers are deepened and altered in hue by increase of thickness, for the same reason as those of translucent bodies. Thus gamboge in thin washes is yellow; in thicker, orange: carmine in thin washes is nearly pink; in thicker, nearly red: Prussian blue in thin washes is seagreen-blue; in thicker, blue.

26. *Inferiority of colours from internal reflection.* The colours of bodies seen in this way cannot equal in depth and richness those which may be derived from transmitted light. Not only is their light in general scattered, and therefore much weaker than that which transparent bodies may transmit when the sun or other brilliant white light is viewed directly

through them ; but they are also necessarily more or
less mixed with the white light of superficial reflection,
which is always considerable, and often so great as to
overpower the peculiar colour derived from the light
internally reflected. Thus the colours of a painting
are not well discernible from a point of view in which
the eye receives also a large quantity of light reflected
from its surface. Hence to see objects of this kind to
best advantage they should be viewed in some direction
very divergent from the direction in which the greatest
quantity of incident light is reflected, so that their
proper colours may be as little as possible diluted.

The effect of a transparent varnish, laid over a
coloured surface which scatters light by a superficial
reflection, is to deepen the proper colour from most
points of view by preventing reflection from the original
surface, and producing on itself a glossy surface which
reflects light regularly, instead of scattering it in all
directions.

27. *Effect of the form and texture of a surface.* The
colours of all such bodies are also much deepened
when the texture or form of the body is such as to
cause repeated reflections from one part of it to another.
Hence the depth of colour which may be seen in vel-
vets and woollen stuffs and in the hollows and interstices
between the petals of flowers, which is lost when the
parts are viewed separately. The effect is the same as
if the object were illuminated by light of its own
proper colour; the superficial reflection being tinged,
as well as the internal reflection being deepened in
colour by the repetition of the process of absorption.
But in general the hue is also altered more or less by
this means, from the cause which modifies the colours
of translucent bodies when their thickness is increased;
namely, the unequal rates of absorption for the different kinds of light.

28. *Colours from superficial reflection.* Thirdly,
colours are produced by light reflected from the outer
surface of an object.

Endless varieties in the appearance of bodies are produced by reflection, according to the form and nature of the surface, the reflecting power of the substance, and the nature and direction of the illumination. In superficial reflection the quantity of light reflected increases considerably with the obliquity of the incident light; but this is more apparent with polished surfaces than with those which scatter the light in all directions. Two kinds of superficial reflection must be noticed.

Ordinary superficial reflection is almost wholly the same for all kinds of light, and therefore dilutes with white the colours produced in other ways. Its effect is best observed in bodies which powerfully absorb all light which enters them, as black marble, rough or polished, where the superficial not being complicated with the internal reflection, it is easy to see that the hue of the reflected light is the same as that of the illuminating light.

29. *Colours from metallic superficial reflection.* Certain bodies, however, especially metals, possess an extraordinary power of superficial reflection, attended with almost total opacity. The reflective power of silver (the most intensely reflective of all bodies) is very nearly equal for all kinds of light; and this is also the case with most metals, but not with all. Thus gold and copper reflect yellow and red light most powerfully.

A few non-metallic substances also have the same property, either for all, or for some kinds of light. Plumbago for instance gives a gray metallic reflection; compressed indigo gives a red one; potassium permanganate a green one (as may be seen in its crystals, and even in a strong solution of it, which is a beautiful pink by transmitted light and very opaque to the green rays).

The splendid colours of some birds and insects, distinguished by their peculiar metallic appearance, and by the richness of their hues increasing when the reflected light increases, instead of being more and more diluted by white, are also due, in part at least, to this kind of superficial reflection.

30. *Effect of repeating reflections.* Colours produced by superficial reflection are deepened and altered in hue by causing the light to be repeatedly reflected from one part of the surface of the body to another. Thus the inside of a gold cup gives a deep orange red, instead of yellow, like a plane surface of the same metal; and even pure silver under like circumstances gives a pale yellow, though on a plane surface it shows no perceptible tendency to any hue.

The colour to which such bodies tend under these conditions is of course the colour of that kind of light which they most copiously reflect.

31. *Colours from reflection from small particles.* Fourthly, colours arise from a peculiar kind of reflection which takes place from small foreign particles included in a transparent medium, whether gaseous liquid or solid.

When the particles are very small the only kinds of light they reflect are those which are most refrangible; and when the particles are larger they reflect more and more of the other kinds also. Hence the colour of the reflection is deep blue in the first case, then becomes more azure and paler, until, when the particles are large enough, it is white; and the light thus reflected is always scattered in all directions.

Hence the beautiful blue of the sky, which is reflected from extremely minute particles of water produced by the condensation of aqueous vapour near the uttermost limit of the atmosphere; larger particles of the same kind in the clouds, reflecting white. The blue of the lower part of the atmosphere, the blue of wood smoke, of diluted milk, and of some kinds of opalescent stones and glass and other natural objects (viewed against a black background), is attributable to like causes. The consequence of such objects scattering so much of the incident blue, or blue and seagreen light, is of course that they appear by transmitted light of a yellow, orange, or reddish hue.*

* See Professor Tyndall on this. Philosophical Transactions, 1869.

32. *Colours from interference.* Fifthly, variable colours are produced in a great variety of ways by what is called the interference of the waves of light, whereby certain kinds of light are destroyed, and other kinds are increased in intensity, on some parts of the surface of a body ; while the reverse takes place on other parts, according to the direction in which the surface is viewed.

The two most common groups included in this class are (1) the colours of finely-striated surfaces, such as those of mother-of-pearl, which have been successfully imitated by engraving lines extremely narrow and close together upon steel or glass ; (2) the colours of thin plates such as those of soap-bubbles, of the wings and scales of some insects and fishes, of the coat of dross that forms on the surface of some metals, and the iridescence often seen on ancient glass. Such colours may be produced in great beauty by reflecting light from the thin layer of air between two pieces of clean polished glass pressed hard together. Looking through the thin layer of air pale colours may also be seen, complementary to those of the reflected light.*

33. *Colours from fluorescence.* Sixthly, colours are produced by what is called fluorescence, a property which certain bodies have of emitting new light of certain kinds when they are acted on by rays of other kinds.

Greenish fluor spar, for instance, when placed in sunshine emits a peculiar blue luminosity, from which the term is derived. The like occurs with a decoction of the bark of the horse-chesnut and of the ash tree, and with a solution of quinine. Uranium glass exhibits a yellow-green fluorescence, and a tincture of the green colouring matter of leaves, a red fluorescence. Objects which have this property may be regarded as self-lumi-

---

* By transmitting what is called polarized light through a plate of mica, selenite, or other substance which possesses the property of double refraction, interference-colours are produced in great beauty, which by making the plate revolve through a quarter of a circle change into their perfect complementaries through gray, as from red to seagreen, green to pink, blue to yellow; but the necessary apparatus is somewhat costly, and the full explanation lengthy.

nous in some degree, under the action of those rays which excite it; especially as these are often invisible rays, more refrangible than the extreme violet rays.[*]

34. *How to test the colours of objects.* The composition of the colours of natural bodies of all sorts may be readily tested by looking through a prism at a stripe of the object continuous with a stripe of white paper of the same breadth, over a perfectly black ground. The rays which the object sends to the eye will appear in their proper places in the spectrum of its stripe, whilst the rays which it does not send, or in which it is defective as compared with white, will be discovered by comparing this spectrum with the adjoining spectrum of the stripe of white paper.

If the colour to be examined belongs to a translucent substance, it may be placed behind part of a slit in an opaque screen, and the light of a white sky be transmitted through it. If it belong to a pigment, a portion of a stripe of white paper may be covered with the pigment (taking care to cover well the edges, and to make the ends square) and the stripe, placed over a dark cavity, may then be viewed through the prism. The colours of flowers may be examined in like manner by laying small rectangular portions of the leaves on a stripe of white paper of the same width.

Such experiments afford an admirable means of instructing the eye to judge correctly of the constitution of all sorts of colours. The method may also be advantageously used by those who seek to improve the manufacture of pigments, or to make good mixtures of pigments. The more nearly the pigment or mixture reflects some continuous parcel of the prismatic rays without diminution, while it totally destroys the rest, the more striking is its colour.

---

* Fluorescence was first explained by Prof. G. G. Stokes (Phil. Trans. 1852). It has been detected in many bodies, but its effect is generally too small to be noticed. Possibly the red tinge in the extreme violet rays of the spectrum may arise from fluorescence of the retina.

CHAPTER III.

INTERMEDIATE COLOURS, AND MEANS BETWEEN TWO OR MORE GIVEN COLOURS.

35. *How to find intermediate colours.* The colours intermediate between one given colour and another may be shown in the following way :—

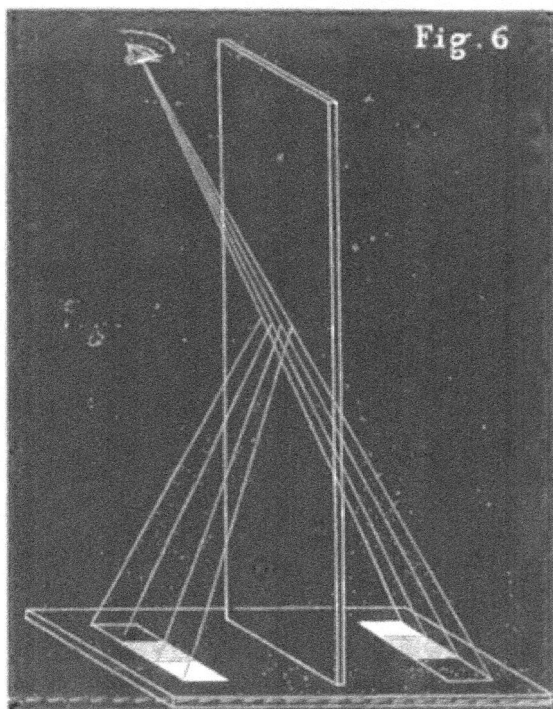

Fig. 6

Lay a spot of each of the two colours upon an horizontal surface, and erect midway between them a slip of clean polished glass as represented in fig. 6

One of the spots may then be seen reflected from that part of the glass through which the other is seen, and their colours are mixed or blended together. When the eye is placed low, the mixture contains most of the colour of the far spot, and very little of the colour of the near spot; when high, the reverse; because the more obliquely the rays fall upon the glass the more they are reflected, and the less they are transmitted. If the eye is raised therefore from a low point to a high one near the plane of the glass, the mixture will change through the whole gradation from the colour of the far spot to that of the near one.

If, instead of spots, stripes of the two colours are laid, meeting each other end to end at the glass, so that the one will be reflected from the part of the glass through which the other is seen, the whole gradation may be viewed at once.

In all these experiments care should be taken that the two spots or stripes are equally illuminated; and it is best to lay them on a neutral ground.

36. *How to find the mean of two colours.* The exact mean between two given colours (that is, the colour which is produced by combining half the light reflected by one of the coloured surfaces with half of that reflected by the other) is easily found by the same method. For, if spaces of white and black are put on either side of spots of the given colours opposite to each other (as shown in the same fig. 6), then where the white reflected by the glass is exactly as strong as the transmitted white, so as to produce the same gray on each side of the blended colours, we may be sure that the resulting colours contain half of each of the given colours.

37. *How to see whether colours are complementary.* If we desire to try whether two colours are perfect complementaries, we have not only to find (1) whether they neutralize each other on the glass in some position of the eye, but (2) whether they do so at that particular position of the eye in which half of each is thrown together, and (3) whether they then produce a gray

equal in darkness to the mean between black and white. The first condition being fulfilled merely shows that their hues are complementary in kind; the second, that they are also of equal strength ; the third alone proves that the lights of the two thrown together would make up the full white, and that they are therefore complementary in light and shade, as well as in hue.

38. *How to find the mean of several colours.* The colour which is the mixture of two, three, or more colours, in any given proportions, may be found by covering a circular disc with the given colours in sectors of the required proportion, and giving it a rapid rotation on its centre in the manner of a top.

39. *How to compare means of different sets of colours.* The mean of a number of given colours may be compared in this way with the mean of any others. Let a circle be drawn midway between the centre and the circumference of the rotating disc, and let the part of the disc outside this circle be covered in the proper proportion with the first set of colours, and the part inside with the second set. When the disc revolves, the means of the two sets will at once be seen in juxtaposition.

Thus, to see whether some given red, green, and blue will neutralize each other to a gray one third of the brightness of white, cover a third of the outer part of the disc with each of them, and let one third of the inner part be white and the rest be black. If the colours are perfectly correct the whole disc when revolving should appear one uniform gray.

40. *Truth of results.* The means of two colours found by the method of rotation (half of the rotating disc being covered with each of them) perfectly agree with the results shown by the slip of glass. Those pairs of colours, also, which are found by experiments made with a slip of glass or a rotating disc to have a gray for their mean colour, always correspond in hue with colours which appear by experiments with the prism to be complementary with each other. Indeed, the more nearly we imitate with pigments the colours

of the prismatic complementaries, the more nearly do the colours of our imitations neutralize each other in equal proportions on the glass or the rotating disc. Thus the colours of vermilion and verdigris, of emerald green and rose madder, of cobalt blue and king's yellow, respectively, which approximate to the standard prismatic primary and secondary colours, produce in either of those ways a nearly perfect gray.

41. *Means between principal colours.* It is useful to notice the effects of mixing thus the principal colours in equal quantities, as in the following instances :—

The means between pairs of full primaries are dark secondaries, the same as the means between the full secondaries and black ; namely, dark seagreen, between green and blue ; dark pink, or crimson-purple, between blue and red ; dark yellow, or olive-green, between red and green.

The means between pairs of full secondaries are light primaries, the same as the means between the full primaries and white ; namely, light red, between pink and yellow ; light green, between yellow and seagreen ; light blue, between seagreen and pink ; which three means are respectively complementary to the former three.

The means between the full primaries and black, and those between the opposite full secondaries and white, are also perfect complementaries ; namely, dark red, and light seagreen ; dark green, and light pink ; dark blue, and light yellow.

The means between the several full primaries and each of their cognate secondaries are all rich hues of medium brightness ; namely, yellow-red, or orange, between red and yellow ; yellow-green, between green and yellow ; seagreen-green, between green and sea-green ; seagreen-blue, or azure, between blue and green ; pink-blue, or puce, between blue and pink ; pink-red, or cerise, between red and pink ; the former three of which are respectively complementary to the latter three.

C

The means between the full primaries and their opposite secondaries are of course all the same gray as the mean between black and white.

The mean or average colour of any set of three primaries of equal strength, is a gray of their mean brightness; and the same is true of any set of three secondaries, and indeed of any set of colours which are equidistant on any chromatic circle, that is, equidistant in hue and equal in strength.

42. *Shades of tints.* Nature abounds with objects whose colours are compounded of the three simple colours in not very unequal quantities; and which therefore do not differ much from a gray. These are called tertiaries, from the character of their composition. Among them may be mentioned maroons, russets, browns, citrines, olive-greens, sage-greens, stone, slate, and lavender colours, in all their endless varieties, light and dark.

All such grayish colours are in fact shades of tints, that is, intermediates between some clear light colour and black; or they may be equally well regarded as tints of shades, that is intermediates between some deep dark colour and white.

43. *How to ascertain the real hues of obscure colours.* The methods suggested for blending colours are not only useful in finding the complementary of any given colour, or the general tone of any composition of colours; they also enable us to detect the real hues of those obscure colours in which the predominance of one or more of the simple sensations over the rest is too small to be easily recognized.

For it is generally easy to see by inspection to which of the six principal gradations the hue of the given colour belongs; that is, whether it lies between pink and red, or red and yellow; between yellow and green, or green and seagreen; or between seagreen and blue, or blue and pink. Let then a stripe of paper, coloured with the gradation to which the hue belongs, be laid parallel to the foot of the vertical slip of glass, and be

reflected against a neutral ground on the other side of
the glass varying in darkness from white to black. It
will be possible so to adjust the position of the eye,
and the distances of the coloured stripe and of the
neutral surface from the glass, that the obscure colour
in question shall be exactly imitated by the mixture
of the strong colour of some part of the stripe with the
dark or light gray of some part of the neutral surface;
and thus the real nature of the colour in question may
be determined.*

CHAPTER IV.

THE NATURAL SYSTEM OF COLOURS AND ITS USE.

44. *Construction of the colour-cube.* Since all colours
are combinations of three simple sensations, any one of
which may vary in intensity from nothing up to the
highest degree which the organ of sight can bear, all
colours may be supposed to be arranged in the form of
a cube in the following manner:—

Let three sides meeting in one corner of the cube be
the sides of no red, no green, no blue, and at right
angles with every part of these sides respectively let
red green and blue be supposed to increase in intensity
up to the strength in which they exist in a full white
in the opposite three sides, which will therefore be the
sides of full red, full green, and full blue. Hence the

---

* Instead of thus dulling a strong colour to match the obscure one,
the object may sometimes be attained by throwing a strong white light
by a lense or reflector upon a spot of the obscure coloured surface, so
strengthening its hue till it matches as nearly as possible the colour of
some part of the less illuminated gradation of colours.

colour belonging to any given point in the cube will be compounded of red green and blue in the proportions of the distances of that point from the three sides of no red, no green, no blue.

Thus if Z in fig. 7 is the corner in which the three initial sides meet, it will be the place of black (the zero of all colour); and if ZGSB is the side of no red, ZBPR the side of no green, and ZRYG the side of no blue, then it easily follows from the construction of the cube that R, G, B, the corners next to Z, will be the places of full red, full green, and full blue, and that S, P, Y, the corners respectively opposite to R, G, B, will be the places of full sea-green, full pink, and full yellow. Also that W, the corner opposite to Z, will be the place of full white; and that the centre being half the distance of W from each of the three initial sides, will be the place of a gray of half the brightness of the full white. In like manner the composition of the colour of any point may be found when its proportionate distances from the initial sides are known.

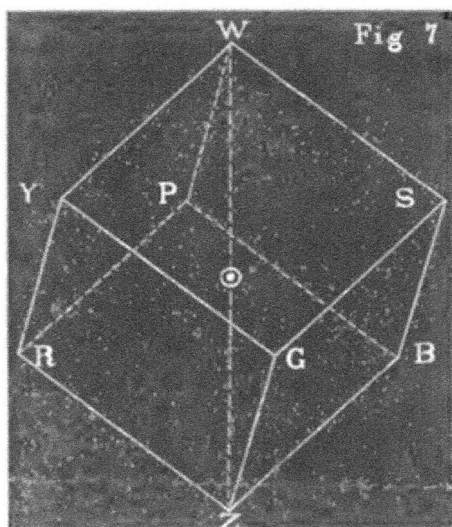

Fig 7

The frontispiece contains two perspective views of the colour-cube, representing the principal colours and their means, twenty-seven in all. The lower diagram shows the three sides adjoining to the corner of black, the upper diagram shows the three opposite sides, reversed in position so that the complementary colours may be vertically over each other. These diagrams should of course be viewed by the white light of day; but even then the failure of the pigments to give a true representation of the colours intended will be very apparent. Some deviate considerably in hue; and their depth and clearness is unequal, though pains have been taken to use the best, and in the most advantageous thickness, so as to give some idea of what the six sides of a perfect colour-cube should be.

45. *Rectilineal gradations.* From the construction of this cube it also follows that every straight line drawn through it, will pass through a uniform gradation of colours from end to end; and also that the gradations along all parallel lines will be of the same nature, being produced by the regular addition or substraction of some one and the same colour. Thus, the gradations along ZR, and all lines parallel to it (fig. 7), are produced by regular increase of red alone; the gradations along ZY, and all lines parallel to it, are produced by the regular and equal increase of both red and green, or of yellow; the gradations along ZW, and all lines parallel to it, are produced by the regular and equal increase of red green and blue, or of white; again, the gradations from R to G, and on all lines parallel to it, are produced by the regular and equal decrease of red and increase of green; and the gradations from R to S, and on all lines parallel to it, are produced by the regular and equal decrease of red and increase of both green and blue.

Hence, the middle point between the places of any two given colours, is the place of their mean colour; and the colours at the extremity of any line passing through the centre of the cube, and bisected by it, are perfect complementaries.

46. *Plane gradations.* Again, the colours that belong to any plane section of the cube will vary according to

certain laws in every direction; and those of all parallel
sections, varying, as they must, in the same manner in
the corresponding directions, will have certain relations
to each other. Thus, the plane ZRYG will present an
orderly group of all colours in which there is no blue;
the parallel plane BPWS will give the same colours
modified by the addition of the full blue to each of
them; and a parallel plane half way between these
two would give them modified by the addition of half
the full blue to each.

Again, any section at right angles with the line
ZW (which may be called the principal axis of the
cube) would present an orderly group of all the colours
possessing some equal degree of brightness or luminosity.

In all sections that can be made through the colour-
cube, the central colour must be the mean or average
of all the colours, both in hue and brightness, and
must therefore express the general tone or effect of the
whole of that peculiar natural arrangement or harmony
of colours which the section presents.

47. *Complementary plane gradations.* If any section
is taken not through the centre of the cube, it is easy
to see that another section may be made parallel to it
at the same distance on the other side of the centre,
which will contain all the colours perfectly complementary to those of the first section.

Thus, a section at right angles with the principal
axis of the cube (ZW) taken near the corner of black,
will pass through the dark primaries and their intermediates; and a corresponding parallel section near the
corner of white will pass through the complementary
light secondaries and their intermediates. Again, the
colours of a section near and even with the edge ZR
will in like manner be complementary to those of the
corresponding section near the edge WS; and so on.

48. *Cognate plane gradations.* Again, as all parts of
a cube are arranged symmetrically on a ternary system
around the principal axis (ZW), any section that is not
taken at right angles with the principal axis may be

accompanied with two other sections, so disposed that the three sections shall all bear the same relation to the three primaries respectively; that is, to the three initial sides.

Thus, ZRWS, the section which contains the red and seagreen colours, is cognate with the sections ZGWP and ZBWY, which contain the green and pink colours, and the blue and yellow colours.

And moreover; if any such section is not symmetrically taken with reference to either of the three initial or terminal sides, it may be accompanied with another section, deviating as much from symmetry as the first does but in the opposite direction; and then, instead of three, there will be three pairs of cognate sections; and they will all represent certain natural arrangements or harmonics of colours having certain relations with each other.

49. *Complementary and cognate rectilineal gradations.* It is obvious too that the regular gradation of colours represented in the cube by any straight line (except the principal axis) must have three or six cognate gradations which may easily be found by noting the places where the line intercepts the sides of the cube; and that all these must have their corresponding complementary gradations; so that sets of gradations may always be found bearing certain relations to each other.

For instance, the following six (see fig. 7) would be cognate gradations :—

From the middle of ZR to the middle of RY;
    From the same to the middle of RP;
From the middle of ZG to the middle of GS;
    From the same to the middle of GY;
From the middle of ZB to the middle of BP;
    From the same to the middle of BS;

(that is, from dark red to orange, from dark red to cerise, from dark green to seagreen-green, from dark green to yellow-green, from dark blue to puce, and from dark blue to azure).

And the following six would be respectively complementary to them :—

From the middle of WS to the middle of SB ;
  From the same to the middle of SG ;
From the middle of WP to the middle of PR ;
  From the same to the middle of PB ;
From the middle of WY to the middle of YG ;
  From the same to the middle of YR ;
(that is, from light seagreen to azure, from light sea-green to seagreen-green, from light pink to cerise, from light pink to puce, from light yellow to yellow-green, and from light yellow to orange).

Also the single colour represented by any given point in the cube (not being in the principal axis) has in like manner its cognate colours, and all these have their complementary colours ; so that regular sets of single colours may also be suggested and easily found by the aid of the colour-cube.

50. *Limits of colours in depth and clearness.* Only colours in which some one or more of the colour-sensations is not at all excited, and which belong therefore to one or other of the three initial sides of the cube, may be regarded as perfect in depth ; while colours in which one or more of the colour-sensations is fully excited, and which therefore belong to one or other of the three terminal sides, may be regarded as perfect in clearness.

In practice, however, it must be remembered that (excepting black and white) we cannot get those theoretical pure or perfect colours which would find their places on the surface of the colour-cube ; but the deeper a colour is, the nearer its place in the cube to one of the three initial sides, and the clearer it is, the nearer to one of the three terminal sides. Hence, as the colours produced by the extreme rays of the red and blue bands of the spectrum are the deepest possible, we may approach nearer to the surface of the cube on the side of no green, to which belong the blues, violets, purples, crimsons, and reds, than to any other initial side. Again, as the perfect complementaries of the deepest colours must be the clearest colours, we may

approach nearer to the opposite side of full green, to
which belong the yellows, greens, and seagreens, than
to any other terminal side. This perfectly accords
with what is observed in the colours of pigments,
flowers, insects, birds, and other natural objects; the
best blues, purples, crimsons, and reds are unsurpassed
in depth, the best yellows, light greens, and seagreens
are unsurpassed in clearness.

51. *Utility of the natural system.* The advantage of
studying the natural system of colour contained in the
colour-cube is not only the infinite variety of different
gradations and arrangements of colour which it suggests,
but also the endless symmetrical combinations of them
which it will enable the designer to invent with a preci-
sion and completeness that otherwise would be impossible; besides the facility with which by its aid the
true affinities of colours may be borne in mind, together
with their practical limits in depth and clearness.

By means of a model, which may be formed by
colouring a small cube of paper, it will be easy to
trace out the colours belonging to any lines and
planes which may be conceived to intersect the cube,
as mentioned in sections 45-49. Thus many beautiful
natural harmonies of colour (as they may be called)
may be studied, besides the six which are represented
in the frontispiece.*

---

* In another treatise, "The Principles of the Science of Colour," this
part of the subject is illustrated by diagrams of natural harmonies of
colours taken from about sixty different sections of the cube, and arranged
so that either each single group of colours, or each set of parallel or of
cognate groups may be viewed together. Methods of forming innumerable
regular compositions out of such natural harmonies, are also suggested
in the same work.

CHAPTER V.

OCULAR MODIFICATIONS OF COLOURS; THE SENSIBILITY OF THE EYE TO LIGHT.

52. *Variable sensibility of the eye.* The excitement of a colour-sensation diminishes for the time the capacity of the retina for the further excitement of that sensation; and any rays of light which then enter the eye do not produce that particular sensation so powerfully as they would do if the eye was not so excited. This defect of sensibility continues for some short period after its cause has ceased to act.

For example, after looking fixedly for some time at a very brilliant red spot against a black background, if we look suddenly at a dark red surface, that part of the retina which received the impression of the bright red will be hardly sensible to the faint red light which it now receives, and a nearly black spot will therefore appear on the dark red surface. Or if in the same case we look at a dark yellow or olive-green surface, a spot of dark green will appear upon it; if we look at a purple surface, a spot of dark blue will appear; if at a gray surface, a spot of dark seagreen: all obvious effects of the same cause, the reduced sensibility of the eye to red light. And if strong green or blue is substituted for the exciting red, and the eye suddenly directed to a surface of some other colour, corresponding effects follow ; an " ocular spectrum," as it is called, being at once seen, whose colour is that which would be produced by subtracting a portion of green or blue from the true colour of the surface we look at.

53. *Law of variation.* The loss of sensibility thus caused is proportionate to the intensity of the excitement ; so that if the three simple colour-sensations are excited all at once in unequal degrees (as in fact they always are when we look at any bright object that is not white), the loss of sensibility will be greatest for the sensation most powerfully excited, and least for the sensation least excited.

Thus if we look at a brilliant light orange surface (such as a bright gold coin) on a dull gray background, and then suddenly remove it without moving the eye, we see in its place a darkish spot on the gray, tinged with the colour complementary to the orange. For the eye which was before most strongly excited to the sensation of red, rather less so to that of yellow, and still less so to that of blue, is now least sensible to the red component of the gray, rather more sensible to the green, and most sensible to the blue ; it therefore sees, instead of gray, a dull seagreen-blue, complementary in hue to the orange.

54. *Simultaneous contrast.* As the eye is ever glancing forwards and backwards from one part of the objects viewed to another, the consequence of the variableness of its sensibility is that the different colours of neighbouring parts always more or less modify each other. For they are sure to be some mixture of the three sensations ; and the proportions of the mixture must become altered according to the nature of the previous excitement of the eye.

These modifications of colour are commonly referred to as the effects of " simultaneous contrast."

55. *Effects of simultaneous contrast.* The general effect of simultaneous contrast is to make the difference of two colours in brightness and hue appear greater than it really is.

Thus a bright white subtracts whiteness from neighbouring objects and causes their colours to appear darker and deeper than they really are, but without altering their hue. In like manner,

Strong seagreen subtracts seagreen ; and therefore dulls another seagreen, and tends to alter

Seagreen-blue to Blue,	Seagreen-green to Green,
Blue to Pink-blue,	Green to Yellow-green,
Pink-blue to Pink,	Yellow-green to Yellow,
Pink to Pink-red,	Yellow to Yellow-red,
Pink-red to Red,	Yellow-red to Red.

It deepens red, and tinges gray with red.

Strong pink subtracts pink ; and therefore dulls another pink, and tends to alter

Pink-red to Red,	Pink-blue to Blue,
Red to Yellow-red,	Blue to Seagreen-blue,
Yellow-red to Yellow,	Seagreen-blue to Seagreen,
Yellow to Yellow-green,	Seagreen to Seagreen-green,
Yellow-green to Green,	Seagreen-green to Green.

It deepens green, and tinges gray with green.

Strong yellow subtracts yellow ; and therefore dulls another yellow, and tends to alter

Yellow-green to Green,	Yellow-red to Red,
Green to Seagreen-green,	Red to Pink-red,
Seagreen-green to Seagreen,	Pink-red to Pink,
Seagreen to Seagreen-blue,	Pink to Pink-blue,
Seagreen-blue to Blue,	Pink-blue to Blue.

It deepens blue, and tinges gray with blue.

Strong red subtracts red ; and therefore dulls another red, and tends to alter

Yellow-red to Yellow,	Pink-red to Pink,
Yellow to Yellow-green,	Pink to Pink-blue,
Yellow-green to Green,	Pink-blue to Blue,
Green to Seagreen-green,	Blue to Seagreen-blue,
Seagreen-green to Seagreen,	Seagreen-blue to Seagreen.

It clears seagreen, and tinges gray with seagreen.

Strong green subtracts green; and therefore dulls another green, and tends to alter

Seagreen-green to Seagreen,	Yellow-green to Yellow,
Seagreen to Seagreen-blue,	Yellow to Yellow-red,
Seagreen-blue to Blue,	Yellow-red to Red,
Blue to Pink-blue,	Red to Pink-red,
Pink-blue to Pink,	Pink-red to Pink.

It clears pink, and tinges gray with pink.

Strong blue subtracts blue; and therefore dulls another blue, and tends to alter

Pink-blue to Pink,	Seagreen-blue to Seagreen,
Pink to Pink-red,	Seagreen to Seagreen-green,
Pink-red to Red,	Seagreen-green to Green,
Red to Yellow-red,	Green to Yellow-green,
Yellow-red to Yellow,	Yellow-Green to Yellow.

It clears yellow, and tinges gray with yellow.

Black of course seems to brighten white, and to make all other colours lighter and clearer; that is, as compared with what they would appear if seen in juxtaposition with any light colour.*

56. *Accommodation of the eye to light.* There is a certain degree of brightness which is on the average most agreeable and suitable to each person's sight. If we take the sensibility of the eye while contemplating the mean of all colours, or the middle gray, belonging to that degree of brightness, as its regular or normal state, there are two ways in which that wonderful organ tends to diminish the difference in the intensity of the sensations that would otherwise arise from a difference in the quantity of light :—

(1) By the diminution of sensibility under increased excitement, and the increase of sensibility under diminished excitement, which has been already discussed in this chapter;

* See Chevreul's treatise on this subject. His chromatic circle, however, and complementary colours require correction from the red-yellow-blue to the red-green-blue system.

(2) By the contraction of the pupil of the eye which always follows an increase of light, and the enlargement which follows a diminution of light. This self-adjustment of the area of the circle that admits light to the eye makes that area under a very bright light less than a tenth part of what it is in the dark.

But though the effects of changes in the illumination of objects are thus in some degree lessened, yet the range of difference in the intensity of the colour-sensations which the eye can receive without injury is extremely large. The brightness of white paper (for example) illuminated by the direct beams of the sun must be hundreds of times greater than its brightness in the shade, even supposing the eye to have accommodated itself to the conditions of each case.

57. *Sensibility of the eye under a given illumination.* With light of a moderate brightness, a good eye is just sensible of an increase or decrease of about one hundredth part of that brightness. This is best shown by illuminating a white screen with two equal lamps, placing one lamp near the screen, and removing the other to such a distance that a shadow cast by it upon the screen first becomes imperceptible (the eye being shaded from the lamps). If the distances of the two lamps from the screen are then measured it will be found that the latter lamp is about ten times further than the former, showing that a difference of less than one hundredth part of the light (namely, of the brightest light by which the eye is at the time affected) is imperceptible.

This rule of course does not hold good when the brightness is so great as to be dazzling to the eye, or so small as to be insufficient for distinct vision. In approaching either of these extremes the eye is necessarily less capable of being sensibly affected by small changes of the intensity of the light. But for a very wide range of brightness, as for instance from that of white paper in sunshine down to that of white paper shaded from the sun and great part of the sky, nearly the same number of steps may be distinguished at once between white and black, the eye being affected by the white in each case and by nothing brighter.

Hence, as soon as the eye has adapted itself to the circumstances in which it is placed it distinguishes shades and hues of colour with nearly equal power under very different degrees of illumination; and the importance is evident of shading the eye from any glaring light, whenever we would discriminate the colours of objects that send nearly the same lights to the eye.*

58. *Number of distinguishable colours.* It would seem then that if the full white with which the eye is affected is of any moderate brightness, about one hundred gradations may be distinguished at once between it and black; and as the eye is no doubt equally sensible to changes in each single colour-sensation, it may be estimated that the whole number of perceptible variations of colour would be about one million (the cube of one hundred), were it possible to excite any one of these sensations by itself to the degree in which they are all excited together in white. But this is so far from being possible that probably not a tenth part of that number of colours could ever be really produced, distinguishable at once under any given intensity of white light.†

59. *Effect of very feeble light.* When the eye is very feebly excited, it seems not to receive exactly the same mixture of sensations from any kind of light as it does under a moderately strong illumination; and the difference is not the same with different kinds of light. For such red and blue surfaces as by full daylight appear equally bright, are no longer so when viewed by a very dim light of the same character. The red

---

* Experiments have been made by Fechner and other continental philosophers on this subject.

† This appears from a comparison of the best observations which have been made on the brightness of the prismatic rays, with their respective strength of hue as determined by Mr. Maxwell.

becomes darker than the blue, as if the one were mixed with black and the other with white. This effect may be best observed if a spot of one colour is put upon the other as a ground, or if spots of both are put near together on a gray ground as nearly as possible equally luminous with them : and as it appears not only when the light is diminished, but when the spots are viewed with nearly closed eyes, it must be an ocular effect. Hence the changes sometimes noticed in the relative brightness of deep reds and blues in a picture under different degrees of illumination. But it is certain that their relative strengths of hue do not change; otherwise what is pure white in the brighter light would no longer be so in the dim light.*

60. *Effect of dazzling light.* On the other hand when the eye is very strongly excited by any kind of light, as by any of the prismatic rays in dazzling brightness, their colours seem all to approach towards white, as if the retina became incapable of receiving the particular sensations which they most powerfully excite in much greater degree than the other sensations, which are doubtless always more or less mixed with the principal ones. Thus the sun's disc viewed through any kind of coloured glass which does not too greatly obscure it, appears nearly white ; more so at least than any white cloud viewed through the same. Yet the glass transmits similar light in both cases, as may be seen by putting white paper in the transmitted sunbeam ; so that it is evident the effect is purely ocular.

---

* In trying these effects care should be taken that the illuminating light varies only in strength; for slight changes in its hue, from altered proportions of its component lights, may considerably affect the brightness of colours in the way mentioned in § 18. Thus the approach of sunset, the clouding over of a blue sky, or even a haze coming on, affects the colours of some pigments in a remarkable manner, by diminishing the proportion of certain prismatic rays in the light which they reflect.

CHAPTER VI.

THE HARMONY OF COLOUR.

61. *Harmony the source of beauty.* When several colour-compositions of like character are compared, whether they consist of different colours or of the same colours differently arranged, it is in general seen at once that some are more pleasing than others. Nor can there be a doubt but that, whether we contemplate works of art, as paintings, mosaics, stained glass, mural decorations, hangings, dresses, furniture, porcelain, jewellery, or natural objects, as the grand scenery of land, sea, and sky, woods, rocks, meadows, plants, leaves, flowers, animals, birds, insects, the vast majority of those who see colours alike are agreed in the objects of their preference in this respect.

Whatever it may be in the association of colours that causes a composition to be thus generally felt to excel in beauty, may be included under the general name of harmony; for the term implies the right fitting together of the colours to produce an agreeable effect.

62. *Principles of harmony.* Colour-compositions are endlessly varied in character. They may have only two or three colours, or they may have many; the colours may meet in sudden contrast, or blend into each other by imperceptible gradations; the dif-

ferent hues may be bordered with other hues, or with white, gray, or black; the colours may be the deepest or the clearest possible, or they may be grayish and nearly neutral in hue; they may be dark or light, or a mixture of all sorts. And as some of all these kinds are found to excel in beauty, it would seem that harmony, the cause of beauty, consists in something independent of all such circumstances.

The following rules, derived from a careful examination of different examples, are founded mainly on the facts (1) that the excitation of each colour-sensation up to a certain degree is attended with pleasure, and also its cessation again after a certain interval; and (2) that when we contemplate a colour-composition we almost unconsciously direct the eye to its different parts in succession, so that that portion of the retina which is endowed with the property of distinct vision (and perhaps every other part as well) becomes affected with or relieved from the different colour-sensations according as the object to which it is directed sends or does not send to the eye the rays which excite such sensation.

63. Rule I. *Each colour-sensation should be excited.* A composition which is complete in itself should affect the eye equally with the three colour-sensations, either jointly or separately. In other words there should be a balance of hue in the whole composition, and its mean colour, or general tone, should be some shade of white.

To illustrate this in the simplest manner, we may compare the effect of filling the whole field of view successively with several grounds of different hues, or neutral, but all of some moderate brightness. At first they all may appear equally pleasing as a change from darkness; but we dwell longest with pleasure on one that is neutral, and the eye is less fatigued when all the three sensations are excited equally to produce the given brightness, than when one or two of them are more strongly excited in excess of the third.

64. RULE II. *Each colour-sensation should be relieved.*
A composition complete in itself should afford relief to
the eye at proper intervals from every colour-sensation
which it excites, either together or separately. Such
relief may be afforded at once, as by the intervention
of black ; or by several steps, as by one or more shades
intervening, or by a regular gradation. The darker
any colour is the less it requires relief.

This may be illustrated in the simplest manner by
comparing the effect of a uniform surface of some
given colour, as red, with that of a surface in which
the same colour is presented in parallel stripes, leaving
equal stripes of black between them. In the first
there is no alteration of excitement and relief; in the
second, alterations are afforded whenever the eye shifts
transversely to the stripes, as it is sure to do ; and the
superiority of the latter is instantly apparent.

Such alternations of any positive colour with black
are always beautiful, whether the colour is (as nearly
as possible) the simple red, green, or blue, or a combination of some two of them, as seagreen, pink, or
yellow, or of all three in pure white, or any mixture
intermediate between these ; and whether the colour
contrasts directly with the black, or changes into it by
distinct steps, or by insensible degrees.

The change from any positive colour to a darker
shade of the same, or to another positive colour which
does not contain the former, must of course be attended
with some measure of relief; but in consequence of
the large mixture of all three sensations in the colours
of natural objects, this relief is far from perfect, except
in such cases as the change from one dark deep primary to another, or to its opposite secondary.

Thus, blue upon white affords relief from the red and
green in the white; yellow upon white affords relief
from the blue in the white; red upon yellow affords
relief from the green in the yellow; dark green upon
bright green affords relief from part of the bright
green ; and so on. Such combinations may have a
pleasing effect though the relief is not perfect ; but if

two colours are very strong in hue and bright they are sure to contain so much in common that the relief is hardly noticeable.

It is obvious that every colour-composition which possesses the advantage of relief must necessarily possess the advantage of excitement with it. *

65. RULE III. *Kindred colours associate well.* Colours nearly related to each other (none of them exciting any one of the colour-sensations in a much higher degree than the others) have always a good effect in juxtaposition.

Thus very dark colours, whatever their hue, are all congruous with black and with each other; very light colours again, whatever their hue, are all congruous with white and with each other; the colours which differ little in hue from the primary red, green, o blue, or the secondary seagreen, pink, or yellow, are congruous respectively with those colours, and with each other; and the colours which differ little from the mean gray, are congruous with it and with each other.

The peculiar beauty of associations of kindred colours may be illustrated by compositions of dark red, dark green, and dark blue upon a black ground, or by compositions of light seagreen, light pink, and light yellow upon a white ground; or again by compositions in which a dark blue, a seagreen-blue, and a pink-blue, appear together; or in which a light yellow, a yellow-red, and a yellow-green, appear together.

Associations of this kind are extremely common in nature. They are seen, for instance, in the various hues and shades of crimson or pink in roses, arising not

---

* It must be remembered that unless a colour-composition is intended to occupy the whole field of view for some length of time, the full observance of this and of the preceding rule is neither needful nor always even desirable. For in such cases the eye may obtain excitement and relief from surrounding objects and the artist is left free to work for any effect he wishes to produce, especially if he can insure the composition being properly accompanied.

only from the different colours of their parts, but from the reflections and shades produced between the petals; in the varieties of blue exhibited by the iris, convolvulus, and larkspur; or of yellow by the daffodil and primrose. They appear in the folds of coloured fabrics; in washes of pigments varying in thickness; in stones; in all sorts of polished woods; in the sky and clouds around the setting sun; and in the charming grayish secondary hues which play on the surface of mother-of-pearl, or on the feathers of doves.

A certain degree of congruity is secured between all colours by the circumstance that is impossible to excite any of the simple sensations of colour in any high degree without a large mixture of the rest.

66. RULE IV. *Light colours with dark associate well.* Colours which differ mainly in light and shade, whatever their hues may be, have always a good effect in juxtaposition, especially if the quantity of unneutralized light is but small in both of them.

Thus complementaries and other colours which differ widely from each other in hue, may often be placed in juxtaposition with great advantage, and mutually improve each other, if these conditions are observed; as may be seen with a dark deep primary and its opposite pale clear secondary, such as french blue and lemon yellow. The most striking contrasts between colours of different hues in flowers are of this kind, as that between purple and yellow in the pansy; between dark red and light orange in the coreopsis, dark blue and lighter crimson in some varieties of the fuchsia, and the dark and light colours of sweet peas and tulips.

But when bright colours that are complementary to each other, or differ widely in hue, are placed together, they have in general a less agreeable effect. There is sometimes a certain harshness in the association of such colours, which is commonly called a discord; as if the eye experienced a difficulty in accommodating itself to the sudden change of one powerful predominant hue into another. When colours of this kind come much together in a composition it is not easy to avoid a gaudy or glaring effect.

67. RULE V. *Gradations should be true and regular.* When a series of colours are so related that the eye in passing from end to end becomes less excited by one or more of the colour-sensations and more excited by the rest, the eye is best pleased when the colours form a true and regular gradation, changing uniformly, or in equal steps. In such a case, the middle colour of the gradation will be the mean between the terminal colours, and each intermediate colour the mean between the colours on either side of it.

The peculiar charm that attends a perfectly true and regular gradation of colour is very remarkable, whether the gradation is from some colour to its complementary, through gray, or merely to black, or to white; or whether it is from one hue to some collateral hue, distant or nearly related. But, except between kindred colours (of which kind some tolerably perfect examples may be found in the vegetable and animal world), they are rare in nature, and require some skill to produce in pigments.

68. RULE VI. *Ends of gradations should be equally striking.* Whenever a single gradation of colours constitutes a colour-composition, or an independent part of one, its extreme colours should be equally striking to the eye, so as to appear to balance or sustain each other.

To do this, it is not necessary that they should be equal either in brightness or in strength of hue; it is sufficient if they seem equal in depth or in clearness, or if the depth of one should seem matched by the clearness of the other. That the peculiar beauty of a balanced gradation is independent of the balance of hue referred to in Rule I., is evident from its being found in gradations in which there is no balance of hue, such as one from a deep red to a clear yellow or to white. That it is independent of contrast with any ground colour, is proved by its appearing whether the gradation is viewed against a perfect black, a pure white, a medium gray, or any coloured ground whatever. Different grounds must, however, affect the application of this rule by inducing ocular modifications of the colours in respect of their depth or clearness.

When the gradation is to be regarded in conjunction with other parts of the composition, the balance necessary to satisfy the eye may be obtained from them, and any incompleteness in itself is not necessarily matter of objection.

It may be added that any intermediate colour in a gradation, that is not the mean between the extreme colours, seems also to require, though in a less degree, the support of a corresponding colour of equal depth and clearness in the opposite part of the gradation; and the observance of this condition enhances the regularity of the gradation.

69. RULE VII. *Contrasting colours should be equally striking.* Two colours opposed to each other, without a gradation between them, and independent of others, should sustain or balance each other in depth and clearness.

This is, in fact, but the extreme case of all those to which the previous rule applies. It will be found fulfilled in many of the most beautifully coloured natural objects, as flowers and birds and insects, in which such contrasts occur.

70. RULE VIII. *Gradations, contrasts, and single colours should correspond.* There should be a correspondence or equivalence between the gradations and contrasts which occur in the different parts of the composition. If there is only one colour which is striking for its depth or clearness, that colour should occupy the middle parts to which the eye is naturally mainly directed, so that it may form balancing gradations and contrasts with all the less striking colours

around it; or else it should surround the less striking colours as a background to them, so as to form the like gradations and contrasts with the less striking colours towards the middle. At any rate it should be disposed so as to secure in some way or other a certain symmetry of colours in the different parts of the composition.

The advantage of this may be easily seen by comparing the effect of a composition in which this rule is attended to with that of another of the same colours with which it is disregarded.

Again, if there are two or more striking colours, they should be disposed so as to balance each other across or around the central parts of the composition, either by one or both of the colours being repeated on each side of the middle, or by one being on one side, and the other on the other.

Landscapes, where the clear colours on the sky are reflected from water in the foreground, afford approximate examples on the former variety; while landscapes, where those clear colours are matched by the deep reds, greens, or other hues, and dark shades of the foreground, do the same for the latter.

71. RULE IX. *Colours and groups of colours should occupy appropriate breadths.* There should be some degree of concordance between the breadths of the spaces occupied by the different colours in the composition, and the breadths which the eye is capable of viewing distinctly and with satisfaction, at the distance from which (for the most part) it is to be viewed, without changing its direction more slowly or more quickly than is usual in contemplating an object.

It is obvious that if the parts which are differently coloured are much too small, they become blended together in the eye into some indistinct average colour, in which the harmony of the different colours is for the most part lost: while if they are much too large, so as to require the eye to move over any considerable angle in passing from one to another, the beauty of

the composition suffers from the inability of the eye
to take in readily the different parts; whence a want
of excitement while it dwells too long on the parts
that are dark, and a want of relief while it dwells on
those that are bright.

Hence in general the breadth may be greater for
colours which do not differ very widely from the medium
gray, than for colours extremely bright or extremely
dark, or having some very powerful hue.

A similar consideration applies to the breadths of
the spaces occupied by groups of kindred colours,
which have on the whole the same sort of effect on
the eye.

72. RULE X. *Variety of colour and treatment may
conduce to beauty.* Variety of colour conduces to the
beauty of a composition, if the conditions of harmony
are not contravened by its introduction. It is easy to
see that increasing the number of different colours,
and thereby increasing also the number of different
kinds of gradations and contrasts, must greatly increase the difficulty of assorting them all so as to
satisfy the eye with their harmony; but if this is
effected equally well, a composition with many colours
will always excel a similar one with few in beauty and
continuing charm.

To illustrate this in the simplest manner, compare
the effect of a pattern of equidistant stripes of gray
upon black, with that of a pattern in which the
equidistant stripes upon black are alternately blue and
yellow (or any other two complementary colours), or
else red, green, and blue (or any other three colours
which will neutralize each other). In each case the
three colour-sensations may be equally excited and
relieved; but in the first there is no variety of colour,
because the three sensations are everywhere combined
in a uniform gray: while in the second there are two
different positive colours, in the third three. The
beauty arising from variety is thus made very apparent.

D

73. *Conditions restrictive of rules.* The application of rules of harmony must of course be restricted by the exigencies and proprieties of the subject of design; its nature, use, importance, value, size, intended distance from the eye, and accompaniments; for all such circumstances must obviously affect our choice, both of colour and of treatment. But besides this, it is frequently required to produce some special effect, congenial with the circumstances of the case; and for this end it is useful to study the mental effects which seem to attend different classes of colour and modes of treatment. These will be recounted in the next chapter.

74. *Necessary compromise of principles.* In attempting to fulfil the laws of harmony in any case, it will be found that a selection or compromise must be made between them. Full conformity with any one of them is usually impossible, much more with all. Did no other difficulty arise, the designer is sure to be hampered more or less by the imperfection of the best material means available for producing any desired effect.

But a very moderate compliance with sound principles produces a sensibly good effect; and if the eye is pleased with some good points in a composition, it will generally be indulgent towards inevitable deficiencies.

It should be remembered, moreover, that while an acquaintance with the principles of harmony, and with the true relations of colours to each other, is most useful, both in designing and in judging of colour-compositions, it does not supersede the advantage of good natural taste and sensibility to colour, or of the skill and facility which can only be acquired by careful practice.

CHAPTER VII.

MENTAL EFFECTS OF COLOURS AND OF COLOUR-COMPOSITIONS.

75. *Impressiveness of colour.* The deepest colours impress the mind most with a certain idea of solemnity and magnificence, which is often intended to be conveyed by such words as strength, intensity, richness, fulness, power. This is the more remarkable because their real brightness and real strength of hue is never great, and may be, as in the deepest black, nothing at all. It would seem that the eye intuitively compares the strength of the predominant colour-sensation with the whole brightness of the colour; and the more nearly these approach to equality, the more impressive ·the colour becomes in comparison with others of the like hue.

Of all positive colours the deepest blue, purple, crimson, and red, have most of this quality; but the deepest green, seagreen-green, azure, and even the deepest orange, yellow-green, and yellow, as in certain rich golden browns and olive-greens, are not quite devoid of it.*

---

* In observing this and other mental effects of colours, it is well to compare all the colours with the medium gray, viewing them against a background of that colour which is the mean or average of all colours.

76. *Liveliness of colour.* Clear colours are also particularly striking, but in a different way. They are best distinguished by their lively effect. Both these and the brighter of the deep colours are often said to be " vivid," and not without reason, since they are apt to look brighter than they really are.

As the clearest possible colours are the perfect complementaries of the deepest, it follows that yellow, light green, seagreen, and the like, next to white, afford the best examples of this class ; yet pink, cerise, and even the best light red and light blue are not without a measure of it.

There is one remarkable difference between these two classes of striking colours ; the deep colours are invariably so, and from their nature always impress the mind as such ; whilst the clear colours lose all their excellence in the presence of some brighter colour. Thus what appears the full white in a composition equally illuminated, seems obscured at once when a sunbeam lights up another part which before was far less brilliant.

77. *Warmth of colour.* The three colour-sensations produce different effects on the mind, or, more probably, produce the same effect in different degrees. Red is usually described as exciting, warm, advancing ; green as refreshing, soft, pleasant ; and blue as quiet, cool, retiring. In truth, perhaps, red has more of an exhilarating or exciting effect (more of what is usually called warmth of colour) than green, and green more than blue.*

To compare the effects of different colours in this respect, they should of course be reduced to an equality in brightness and strength ; for the eye, as in other cases, seems to estimate the effect in proportion to the whole brightness of the colour. Thus, white, though

* It is worthy of notice that this may arise from the fact that red light has the slowest vibrations, and blue the quickest; since a similar difference may be observed in musical notes of slow and quick vibrations ; bass notes being more exciting, and appearing nearer at hand, than treble notes.

containing the sensation of red in a higher degree than the colour of the strongest red pigment, does not seem to equal it in warmth.

If the cube of colours be divided into two equal parts by a plane passing through the corners of black, green, white, and pink, the colours on the red and yellow side of that plane may be regarded as warm, the more so the nearer to red; while those on the blue and seagreen side will be cool (that is, less warm than the average), and the more so the nearer to blue; and the colours in or near the dividing plane have a medium effect, and may be considered indifferent as to warmth or coolness.

The combination of this warmth of hue with depth of colour gives an attractive richness of which cool deep colours are destitute; and its combination with clearness of colour produces a gayness in the light clear reds and yellows, very different from the modest retiring liveliness of the light clear blues and sea-greens.

78. *Effects of colours in colour-compositions.* Compositions partake of the character of their predominant colours. When a variety of colours of like character are brought together, the general result is to make that character more marked: and the effect is further enhanced when the colours are so arranged that they mutually improve each other.

Thus a good composition of the deep primaries with black and with some of their deep intermediates, or of the clear secondaries with white and with some of their clear intermediates, is more impressive, or more lively, than any single colour could be; and the effect is more abiding, because the variety leads the eye to contemplate it longer with pleasure. So also a combination of red, orange, yellow, and other warm hues, dark and light, produces a stronger effect in that direction than any single colour; and the same may be said of a combination of all sorts of cool retiring colours.

So also compositions of dull and nearly neutral

colours are uninteresting, and if the colours are dark, even gloomy and sad, unless redeemed in some degree by the manner in which they are put together.

On the other hand, colours widely differing in character may form compositions extremely beautiful, presenting admirable examples of the harmony of colours, though neither very lively nor very impressive, very warm nor very cool.

79. *Effects of modes of composition.* The character of a composition is also affected by the number of its colours and by the manner in which they are put together.

If the colours are few, the composition in that respect is more simple ; and simplicity itself has a charm where other causes of beauty are present ; but without them is mere poverty. If the colours are many, there is danger lest the advantage of variety be lost in complication, confusion, and superfluity.

Sharp contrasts tend to give strength and vigour ; easy gradations lead to gentle and soft effects. The danger of the former is to degenerate into harshness, of the latter into tameness and weakness. The one may be considered as analogous to rectilineal figures, having angular points or cross lines, the other to figures formed by flowing curves.

80. *Imaginary mental effects.* A feeling of congruity or incongruity which attends the use of colours in certain cases seems to deserve notice here, though perhaps it mainly rises from a mere association of ideas.

When green is used in apparel, for instance, the effect is generally felt to be unpleasant, the preference being given to colours in which red and blue predominate, or at least are combined in equal strength with green. It is true that greens generally are poorer colours than reds and blues of the same brightness ; but this difference is insufficient to account for the objection. It more probably arises from the felt in-

congruity of clothing a human being in a colour
which is so general in vegetation and so uncommon
in animated nature ; and it may be noticed that
greens which differ most from those of plants (either
in depth or clearness or in tendency to seagreen) are
most preferred in dress.

There are many ideas connected with colours by
association, especially by imaginative persons, which
have nothing to do with any mental emotion really
excited by the colour itself. Thus one says that red
on black is "an atrocious combination, suggestive of
the executioner," while another may just as reasonably
ascribe the like effect to a dingy blue, the colour of a
butcher's shirt. One, again, who has his mind im-
pressed with the splendour of dark gold will call the
same colour magnificent, which another who thinks of
bile, considers a sickly yellow.

It is quite common for those who have imbibed
prejudices in favour of certain artificial combinations
of colours to be disgusted when they see something
new or unexpected, though such as have learnt to
judge according to natural taste, and the real effect
on the eye, may be delighted with the same.

Colours of any strong hue, again, have been incon-
siderately condemned in general by some persons, as
unpleasantly garish, or unbecomingly obtrusive; and
colours nearly neutral have obtained an undue preference with many. The cause of this may be traced in
great measure to the unskilful manner in which
colours of strong hue have been applied, in ignorance
of the principles of harmony ; but it may be confi-
dently asserted of every colour, without exception,
that there are circumstances in which it would not
only be beautiful, but the most beautiful of all colours,
as conducing most to the good effect of the whole
composition.

CHAPTER VIII.

PECULIAR AND DEFECTIVE COLOUR-VISION.

81. *Dichromism.* Many persons see only two distinct hues in the prismatic spectrum, even though they are sensible to all its rays. All that part which extends from the beginning of the red to the end of the green band (Sect. 7) appears to them of one hue, beginning dark and deep, and becoming brighter and then paler, till at the verge between the green and blue bands the light appears neutral (the same as a mixture of all the rays). The other hue then begins pale and light, and becomes darker and deeper to the end, extending over all that part of the spectrum which contitutes the blue band. There is every reason to believe that these two hues of dichromic vision are the same as the complementary yellow and blue of ordinary vision.

82. *Defective sensibility to red light.* Defective sensibility to the less refrangible or red rays is not uncommon in persons who can distinguish red as well as other hues, and therefore have the ordinary trichromic vision. The existence of the defect is shown by a portion of the red end of the spectrum being invisible to persons who have it. It must affect more or less in brightness and hue the colours of all objects, except perfect black, by diminishing the quantity of red in their composition, just as

they would be affected in ordinary vision, if illuminated by light that has passed through a sea-green
glass.   But whatever is the colour of those objects
which reflect all the kinds of light that such persons
see, they no doubt judge it to be white, as it must be
the brightest of all their colours, and as objects white
to others must appear of that colour to them.   So
objects white by daylight are commonly judged to be
white by candlelight also, though really yellow, by
reason of the deficiency of blue light.

Hence objects which in ordinary vision are red,
orange, yellow, yellow-green, or green in hue, are all
yellows to such persons, differing only in strength
of hue and brightness ; and those which are seagreen-
blue or pink-blue, are all blues differing only in
strength of hue or in brightness ; while sea-green and
pink are both neutral grays.

83. *Dichromism with defective sensibility.*   This dichromic vision is usually attended with some (and often
with a very great) defect of sensibility for the less
refrangible kinds of light, making objects that send
out chiefly those kinds of light very dark, and modify-
ing the colours of all.   Dark reds, in such cases,
are hardly distinguished from black ; bright reds give
the same shades of yellow as browns and olive greens ;
pink appears blue, and cerise appears gray.*

84. *Effects of colour-designs in such cases.*   The
principles of the ocular modifications of colours, of
their harmony, and mental effects, are of course the
same in dichromic and defective, as in ordinary colour-
vision ; but their application is obviously simplified
when only two hues are distinguished.

The only compositions that can appear alike in
ordinary and dichromic vision, are those containing no
other colour but whites, grays, and blacks, with
yellows and blues in all their varieties of brightness
and of strength of hue ; the colours belonging to the
plane which passes through the corners of black, blue,

* Other peculiarities of colour-vision besides these are sometimes
found, but rarely.

white, and yellow in the colour-cube. Compositions of green and blue, with their intermediate hues of seagreen, will appear in dichromic vision as equally regular compositions of yellow and blue and their intermediates. The same may be said of compositions of blue and red, with the intermediate crimsons, pinks, and mauves, when there is no defect of sensibility to the less refrangible rays.

Compositions on the other hand in which reds, greens, crimsons, or seagreens prevail, however beautiful to the ordinary eye, will be generally poor and deficient in variety to persons of dichromic vision, nor will it be possible to them to see any reason in the assortment of the colours adopted.

As a general rule therefore to make colour-compositions pleasing to persons of dichromic and defective colour-vision as well as to others, they should contain only the colours which are the same to each; and if bright red, orange, yellow-green, or green, is put in, it should only be where a yellow of corresponding strength would have a good effect; if seagreen-blue, pink, or mauve is put in, it should be where a blue of corresponding strength would look well; and if crimson or seagreen is put in, it should be where a gray of proper darkness might be substituted without objection; and dark red where black might take its place.

In many cases a system of colouring may well be adopted which would be acceptable to persons who have these very common peculiarities of vision, without displeasing others.